The Complete Guide To Investing in Gold and Silver

Surviving The Great Economic Depression

Copyright Notices

Copyright © 2012 by Omar Johnson
All rights reserved.

No part of this publication may be reproduced or transmitted in any form or by any means, mechanical or electronic, including photocopying and recording, or by any information storage and retrieval system, without permission in writing from the publisher. Requests for permission for further information should be addressed to

Omar Johnson
Make Profits Easy LLC
497 West Side Avenue
Suite 134
Jersey City, N.J. 07304

This book is dedicated to my father.
Thank you for inspiring me.

Legal Notices

While all attempts have been made to verify information provided in this publication, neither the author nor the Publisher assumes any responsibility for errors, omissions or contrary interpretation of the subject matter herein.

This publication is not intended for use as a source of legal, investment or accounting advice. The Publisher wants to stress that the information contained herein may be subject to varying state and/or local laws or regulations. All users are advised to retain competent counsel to determine what state and/or local laws or regulations may apply to the user's particular business.

The purchaser or reader of this publication assumes responsibility for the use of these materials and information. Adherence to all applicable laws and regulations, both federal and state and local, governing professional licensing, business practices, advertising and all other aspects of doing business in United States or any other jurisdiction is the sole responsibility of the purchaser or reader. The author and Publisher assume no responsibility or liability whatsoever on the behalf of any purchaser or reader of these materials. We expressly do not guarantee any results you may or may not get as a result of following our recommendations. You must test everything for yourself. Any perceived slights of specific people or organizations is unintentional.

Table of Contents

Introduction 7

How The Federal Reserve Works 10

Fractional Reserve Lending 13

Understanding Inflation, Deflation, Stagflation and Hyperinflation 16

Currency Vs Money 25

The Barter System 27

Gold and Silver First Uses As Money 28

The Rise And Fall of Athens 29

The Roman Republic 31

The Tulip Mania- A Lesson In Absurdity 35

John Law and The Paper Currency Ponzi 38

Whatever Happened To The Classical Gold Standard? 46

How The Federal Reserve Was Manipulated Into Existence . 48

The International Bankers Fund Both Sides In A War 53

The Fake Gold Standard 55

The Roaring 20's Boom Then Bust 58

The Great Depression 60

Bank Holiday 62

Did FDR's New Deal Get Us Out Of The Depression? 65

Bretton Woods Agreement 67

Gold Made Legal Again! 70

Inflation Comes Roaring Back 73

The Stock Market Crash Of 1987 75

The Dotcom Bubble Then Burst 78

The Real Estate Bubble Then Burst And The Sinking Of The Economy 80

The National Debt 84

To Save or To Hedge 89

Silver and Gold Price Manipulation 94

How To Invest In Gold and Silver 96

I.R.S. Reporting Requirements 102

There are 4 ways to own future contacts 109

Mining stocks 111

Conclusion 111

Other Books Written By Omar Johnson 115

Introduction

Hello my name is Omar Johnson and welcome to **"The Complete Guide To Investing In Gold And Silver: Surviving The Great Economic Depression.** Let me start off by saying we are living in a very dangerous time period. There is a complete global economic meltdown and if you are a citizen of the United States Of America it is imperative that you understand that we are at the forefront of this economic crisis primarily because we've borrowed and spent too much money as consumers and as a nation. We are relying heavily on foreign countries to lend us money to subsidize our standard of living.

As a result of this irrational excessiveness in a few short years we have gone from being a creditor nation to being the largest debtor nation the world has ever seen. In addition, the taxpaying citizens have been forced to participate with the government to bail out the auto industry, failing banks and financial institutions such as AIG supposedly because they are "too big to fail".

It is in my opinion that if we truly have a free market system no one institution is too big to fail. I say if these companies are not making sound prudent financial decisions let them fail and let the so-called free market system work! That's just good economics and as famed investor Jim Rogers noted "when companies fail the way it suppose to work is the competent people take over the assets of the incompetent people".

I truly agree with him in that respect and I'll also add that if the market is not allowed to work this way without government interference what happens is the people and the companies who are prudent and have made sound business decisions get penalized and are told to compete fairly with companies that have received enormous bail out money which is simply ludicrous and distorts real economics.

You are probably wondering at this point what does this have to do with investing in gold and silver? The answer is when these companies are being bailed out that money has to come from The Federal Reserve who most people mistakenly believe is part of the United States Government. The Federal Reserve is neither federal nor does it has any reserves and is a private cartel made up of international bankers who own the nation's money supply.

How The Federal Reserve Works

That's right you heard right the United States of America doesn't even own its' own money supply. It has to go to The Federal Reserve for money. It works like this. Say the government needed 20 billion dollars. It goes to the Federal Reserve and the Federal Reserve buys 20 billion in government bonds in the form of treasuries and then the government sends these treasuries over to the Federal Reserve who in turn issues dollars to the government.

These dollars are called federal reserve notes. That's why if you look on your dollar bills they have the words federal reserve note typed across them. A note simply means that you owe somebody. For example, your car note means that you owe the bank the principal and the interest for loaning you the money to purchase your car. The same principle applies to the mortgage on your house. It is known as a note.

So the money borrowed from the Federal Reserve has to be paid back with interest. And who pays the Federal Reserve back? You guessed it the American people in the form of taxes. That's why after the Federal Reserve bank was created in 1913 the Internal Revenue Service was created in 1914 to make sure and guarantee that the citizens of the United States pay the bankers back the enormous amount of money that the government borrowed to fund wars, build bridges and roads and create social programs and services.

To continue, this 20 billion dollars or Federal Reserve notes are then deposited in the bank and 20 billion dollars is added to the money supply out of thin air. Of course in actual reality this is all done electronically. So in essence since these Federal Reserve notes or dollars are only created out of debt, all money in circulation could not exist without debt so therefore **money = debt.**

Also you must realize that when the Federal Reserve prints money it devalues the money already in circulation. So as the Federal Reserve prints more and more money as in the case of bailouts and economic stimulus packages they debase the currency and this has severe economic consequences such as inflation or rising prices.

In addition, the people who work hard and save their money in a bank become the ultimate losers because they get taxed for earning interest on the money that they save and the low rate of interest that the bank offers them to deposit their money doesn't keep up with the rate of inflation.

Ultimately they get wiped out to because the Federal Reserve just keeps printing more and more money and they can't save as fast as the Federal Reserve can run the printing press.

Fractional Reserve Lending

They call what Bernie Madoff did the biggest ponzi scheme of all time. I beg to differ. The biggest ponzi scheme that was ever created was the Federal Reserve cartel and the fractional reserve banking system. Here's why. The money that the Federal Reserve and the banks create out of thin air is backed by nothing. It was once backed by gold until President Nixon took us off the gold standard in 1971. The dollar was actually redeemable in gold.

Once the United States went off the gold standard the dollar essentially became a fiat currency. What is a fiat currency? A fiat currency is basically money that is back by nothing. Here is an example of how the fractional reserve bank lending system works. When the 20 billion dollars that we've previously discussed is deposited in a commercial bank by the United States government that 20 billion instantly becomes part of that bank reserves.

Under fraction reserve lending requirements in most cases that bank has to keep 10% or 2 billion on hand as a reserve requirement. The remaining 18 billion or 90% is considered excessive reserves and the bank is allowed to use this as basis for issuing loans.

However, this is where it becomes even more ponziesk. You would think that the 18 billion dollars that the bank is using as a basis for issuing loans is coming out of the original 20 billion dollar deposit but it is not! It is created out of thin air because the bank doesn't pay out loans from the money they receive as deposits.

What they do is when they make loans they accept promissory notes and credit the borrower's bank account for the amount of the loan. When that loan is deposited in the borrower's bank account the same fractional reserve process just explained continues again with the borrower's bank thus further expanding the money supply.

To reiterate, the bottom line to all this madness is once again as more dollars are pumped into circulation this inevitably leads to more inflation which leads to higher prices in the marketplace and a decrease in the purchasing power of the dollar.

In addition, when the Federal Reserve who also controls interest rates keeps interest rates low they create bubbles that eventually burst as we have seen with the dotcom and real estate bubbles and bubble bursts which we will discuss the ramifications of.

So it is imperative that you know that the all mighty Federal Reserve is responsible for creating and engineering the economic cycle. All recessions, depressions, recoveries, prosperity, deflation, inflation, stagflation, hyperinflation are all created by The Federal Reserve monetary policies and its' ability to set interest rates and expand or contract the amount of money in circulation at their discretion.

Understanding Inflation, Deflation, Stagflation and Hyperinflation

To understand the reasons why you should be investing in gold and silver now it is necessary for you to comprehend the concepts of inflation, deflation, stagflation and hyperinflation.

Inflation

Let's start with inflation. The word inflate in its' simplest form means to expand and inflation is basically the result of when the money supply and credit are expanded. When there is inflation, prices rise because there is too much capital chasing too few opportunities. It is important to remember that the value of the dollar does not stay constant when there is inflation.

For example, let's say the inflation rate is 10% annually and a sack of potatoes cost $5. In a year that sack of potatoes goes up to $5.50. So after inflation your dollars can't buy the same goods it could beforehand.

The way that the government measures inflation is by using the consumer price index known as the CPI for short and the producer price index or PPI.

According to the government the consumer price index is supposed to be a measure of price changes in consumer goods and services such as gasoline, food, clothing and automobiles. The CPI also supposed to measure price change from the perspective of the purchaser, but these figures that are produced by the department of labor are inherently flawed.

The same thing goes for the producer price index which supposed to measure the average change over time in selling price by domestic producers of goods and services. The PPI according to the government measures price change from the perspective of the seller. The CPI and PPI measurements are flawed because they are constructed in a way that understate price increases and they ignore inflation that are not reflected in consumer or producer prices.

For example, the CPI missed the housing bubble and burst because from 2002-2006 the CPI showed only an 11% rise in home ownership costs while the National Association Of Realtors during this time period reported that home prices soared 34%. The CPI figures were grossly understated because it looks at equivalent rents rather than home prices.

The bottom line is these two indexes that the government uses to measure inflation are inadequate, manipulated, creatively adjusted and consists of all kind of gimmicks which distorts the true picture of inflation.

In addition to this, it is also standard practice at the Federal Reserve under the chairmanship of Alan Greenspan and Ben Bernanke to add liquidity to the economy by increasing the money supply which results in invisible inflation that is not taken into account when inflation is measured.

Deflation

Deflation is simply a contraction of the money supply or credit. In deflation you have declining prices which happened during the last great depression. During this depression the stock market crashed, banks went belly up, there was rampant foreclosure and mega deflation set in. Prices on everything became much less as businesses dumped inventory to try to stay afloat. During this deflationary time the U.S. economy lost 1/3 of its currency supply.

Stagflation

Stagflation occurs when there is high inflation and high unemployment occurring simultaneously. This happened in the United States in the 1970's when the economy wasn't growing and was stagnant but the prices were rising, thus the term stagflation was coined.

Hyperinflation And The Weimar Republic

Hyperinflation is super rapid inflation in which prices increase so quickly that money loses its convenience as a medium of exchange. A classic example of hyperinflation is what happened to Weimar Republic in Germany. At the beginning of World War I Germany went off the gold standard and began running the printing press to pay for the war. They also suspended the rights to redeem the mark which was the country's currency for gold and silver.

The running of the printing press non-stop had the effect of quadrupling the amount of marks in circulation. Although there was a tremendous increase in the money supply, prices didn't rise immediately to keep pace with the inflation as it should have. The main reason for this was people were uncertain about the war so they held on and saved their money.

However, when the war ended confidence was back and people begin to spend their money and this resulted in rapidly rising prices to catch up with the previously created inflation.

In fact, at one stage during this time gold jumped from 100 marks per ounce to as high as 2000 marks per ounce. The people who were smart enough to buy gold when it was 100 marks per ounce of course made out like bandits and preserved, protected and multiplied their wealth while those who unfortunately kept and saved their money in marks saw that their money bought them less and less. In fact in 1919, their money bought them a whopping 90% less when inflation spiraled upward.

For a short time period inflation actually cooled off, but this didn't last for long because World War I was now over and Germany signed the Treaty of Versailles, which placed responsibility for the war on Germany. This was not good for the German economy because it was forced to pay reparations to France, so the German government accelerated the printing of money even more.

Prices started to rise again and eventually they rose 700%. The citizens were disgusted and tried to get rid of their currency as soon as they got it in their hands. Nobody wanted to hold on to their German marks.

Meanwhile, Germany made its' first reparation payment to France mostly with the gold that it had along with other natural resources such as iron, wood, coal etc. However, Germany couldn't make the second payment that was due. In July 1920, the German mark plunged dramatically as the Weimar government informed the Allies it could not meet the schedule of payments, but that it would continue disbursements of coal and other natural resources.

France's premier Raymond Poincare accused Germany of deliberately withholding payments and trying to force the Allies to make concessions by ruining its own currency. France thought Germany was trying to get out of making its payments all together. So on January 11, 1923 French and Belgian troops occupied the Ruhr, a region which furnished 4/5's of Germany's coal and steel production.

The German Weimar government in response to this takeover instructed the factory workers not to work for the enemy and they printed more money to pay them not to, allowing inflation to spiral completely out of control. The mark was becoming worthless as the government printed more and more money.

Just to give you an idea of how much money the German government was printing by late 1923, the German government required 1,783 printing presses, running around the clock, to print money. Despite their Herculean efforts, the economy was collapsing.

To give you an illustration on how worthless the German mark had become, Germans wheeled shopping carts filled with literally trillions of marks to pay for a single loaf of bread. Employees asked to be paid their wages each morning so that they could shop at noon before merchants posted the afternoon price rises.

The only thing that outpaced inflation in Weimar was Gold and Silver. As I indicated earlier gold in Weimar right before the war ended was 100 marks an ounce and during the insidious period of hyperinflation jumped all the way to 87 trillion marks per ounce! Why was this so? Because Gold and Silver are real money and will always outlive and outperform fiat currency which is paper money that is back by nothing.

I have given you this discourse on the hyperinflation that existed in Weimar Germany, because it is relevant history compared to what's going on today as Central banks around the world continue this policy of printing money backed by nothing and debasing their currency.

Many other lessons can also be learned from history regarding inflation, deflation, stagflation and hyperinflation. In fact, it behooves you to learn from history so that you can know how to maneuver in the present as well as the future because as the saying goes "history repeats itself" and those who don't study history often become a victim of it.

Currency Vs Money

Most people think that the dollars that have in their possession or in a bank is actually money but it is not. It is simply just a currency that is used as a medium of exchange to acquire things.

A currency does not store value in and of itself so it is not real money. Real money on the other hand stores value in of itself and that's why gold and silver are considered real money. Conversely, gold and silver can never go to zero in value, but paper money back by nothing historically steadily declines in purchasing power over time and eventually does go to its intrinsic value of zero. The United States dollar will be no different.

Like I said before you must study and pay attention to past history to predict future trends. For example, to finance the Revolutionary war the continental congress issued a currency called the continental. This currency was backed by nothing except the anticipation of taxes that might be collected in the future if the colonies won.

However, most farmers and merchants would not accept this form of currency in exchange for food and goods and George Washington's men often went hungry. The continental was worthless, which led to the popular saying "not worth a continental" meaning that something was as worthless as that paper money.

Now that you understand the difference between a currency and real money let's move on to discuss the history of gold and silver and their relationship to fiat currency, the engineered economic cycles and the various booms and busts that have occurred over time.

The Barter System

Early man's commerce was built on the barter system which involved a simple exchange of goods. For example, a craftsman would exchange a tanned hide for a supply of grain from a farmer. However, the barter system presented many challenges and depended upon double coincidence meaning that the craftsman had to hope that the farmer needed a tanned hide so that he could exchange it for a supply of grain and the farmer hoped that his supply of grain would get him a tanned hide from the craftsman.

In addition, the other problem that the barter system presented was that most goods couldn't be divided. For example, a person who made a sheepskin coat with the hopes of exchanging it for two sacks of potatoes couldn't cut that sheepskin coat in half to receive one sack of potatoes. The whole barter system was limited and had many constraints. So eventually, the whole barter system evolved to the use of precious metals as a medium of exchange.

Gold and Silver First Uses As Money

Although gold and silver were used for over 4,000 years as a medium of exchange, they first became money in Lydia which was located in the southwestern part of modern day Turkey in 560 B.C. The way the story goes is the Lydians had an abundance of electrum which is a naturally occurring alloy of gold and silver. Gold and silver by themselves were already being used as a means of exchange.

However, the Lydians wanted to make use of the electrum so they minted them into coins called a stater and they stamped these coins with a lion's head which was the symbol of the King. Stater coins were made in denominations of 1/3, 1/6 and 1/96. The value of these coins centered on the actual weight of the coin.

The Rise And Fall of Athens

The minting of gold and silver coins progressed to other civilization such as the one in Athens. The Athenians actually thrived initially because they engaged in a free market system and they also had an adequate and efficient way of collecting taxes. Athens was the world's first democracy and it built most of its wealth primarily through trading.

Athens also had very productive silver mines and a progressive government which used the wealth it garnered from gold, silver and trading to build great public works such as the Parthenon and other great buildings of the Acropolis. Athens also used its wealth to fund the many wars that it engaged in as well as its' defense.

However, these costly expenditures became Athens downfall primarily because these wars lasted a long time as in the case of the Peloponnesian war which lasted for 27 years. By 407 B.C. the amount of gold and silver dwindled substantially and the Athenians response to this was to debase the currency by incorporating a significant amount of copper in it as a way to continue funding the war.

Eventually this currency was rendered valueless and Athens was unable to fund the war and lost. A great civilization was destroyed because of its' greed, penchant for war, overspending on public works and a debasement of its currency.

The Roman Republic

The Roman empire replaced the Greek empire and became the dominant civilization. The Romans were a beneficiary of past lessons regarding the abuse and regulation of money from other empires, but they never fully learned from them.

Once again history repeated itself with Rome entering in costly wars and debasing its currency to fund those wars. This ultimately led to an inflated money supply and incredible price inflation.

The way the debasement occurred were that coins were made smaller or coins were clipped. Clipping entailed shaving a small portion off the edges of the coins as a tax when entering in government buildings. These clippings were then melted down with tin to produce more coins. Also, the Romans like the Greeks began to mix their gold and silver with copper and other metals.

Lastly, the Romans invented a practice called revaluation, which was the practice of minting the same coins as before, but stamping them with a higher value. As a result of these practices, inflation spiraled out of control and the Roman populace was highly upset.

In 301 A.D., the ascender to the throne Diocletian issued the Edict of Prices which was an attempt to control runaway inflation and poverty in the Roman Empire. The penalty imposed for those who exceeded the prices of the Edict was death. Not satisfied to execute just the seller, Diocletian decreed that the buyer was to be executed as well. Wages were also frozen.

However, in spite of the Edict prices still kept rising. The Roman currency had become next to worthless and the economy had started to unwind. Rome had a population at this time of 1 million and a great deal of the population was broke, hungry and up in arms.

In fact, the Roman government was forced to hand out free wheat to approximately 200,000 citizens. The Romans were the very first to invent the welfare system we presently have today. The Roman economy was so impoverished that Diocletian adopted a gun and butter policy by putting people to work by hiring new soldiers and by funding public works projects.

This substantially increased the role and the size of government and also tremendously increased deficit spending which is defined as the amount by which a government, company or individual's spending exceeds it income over a period of time.

Rome continued to further debase its' currency and the aforementioned combining factors led to the first documented case of hyperinflation ever recorded.

In 301 A.D. Diocletian specified that a pound of gold was worth 50,000 denari, but the market rate deteriorated to 100,000 denari to buy one pound of gold in 307 A.D. This price incredibly jumped even further to 2.1 billion denari to buy one pound of gold in the middle of the 4th Century!

Once again in the annals of history gold and silver won out and fiat currency was rendered worthless. Needless to say Rome was completely destroyed. Hopefully at this point

you are starting to see the
parallels to what's happening today.

The Tulip Mania- A Lesson In Absurdity

The reason why I am covering this part of history is to show the absurdity of how people are willing to obscenely speculate in the market place causing great economic bubbles that eventually burst like they did with the dotcom era, the stock market's rise and collapse and the recent real estate bubble burst debacle.

I don't know about you but a tulip to me is just a flower. And a flower is well, just a flower. That's how I basically see it. There's no big deal. However, this wasn't the view that was shared by the Dutch in Amsterdam. In 1583, the tulip was actually imported from Turkey and it instantly became a status symbol for the wealthy people of Holland during that time. The novelty of the tulip made it very pricey.

What made it even more pricier was the tulip contracted a non-fatal virus called a mosaic which actually didn't kill the tulip population, but altered them causing flames of color to appear on the petals. The color patterns came in a wide variety which made the tulip even rarer and unique.

As a result of this new dimension to the tulip, prices began to rise even higher. Everyone began to deal in tulip bulbs essentially speculating on the tulip market. This ultimately became a mania and a tulip exchange was established in Amsterdam. This mania eventually led to an economic bubble.

But wait a minute! The absurdity doesn't stop there! The prices of the tulip started rising even faster and people began trading their land, life savings and anything else of value so they could liquidate and buy more tulip bulbs. I guess they figured they would hoard the tulips and sell them for an even higher price to the remaining suckers in the pursuit of enormous profits. Of course the prices that the tulips were selling for weren't a reflection of their actual value.

As it goes in the way of speculative bubbles the wise and smart began to sell and cash in on their profits. This had a domino effect and everyone tried to unload the tulip as prices came crashing down. The problem with this was not many people were buying. This caused people to panic and sell the tulip for whatever price that they could get for them regardless of their losses.

Dealers started to refuse to honor contracts and people woke up to the fact that they traded their homes, life savings and valuables for a flower! The government tried to step in to minimize the losses sounds familiar? If it doesn't refer to the housing crisis where the government of the United States is offering programs to those homeowners who houses are underwater which basically means they owe more on the house than what the house is worth.

The Dutch government intervention in the attempt to halt the tulip crash consisted of offering to honor contracts at 10% of the face value. However, the market plunged even lower making their attempts at restitution impossible. The game was over, leaving in its wake tremendous financial devastation that was felt by the Dutch for years to come.

John Law and The Paper Currency Ponzi

John Law was a Scottish man and the son of a successful banker and goldsmith and was known for his prowess with the ladies. In fact, it was this prowess that stirred up a fight between him and another man over a woman. Back in those days disputes like these were settle over a duel. When Law was challenged to a duel he accepted and shot and killed his challenger.

He was arrested and sentenced to hang, but somehow using his wits he escaped this fate and fled to France. During this time under the reign of Louis XIV, France had a huge national debt. It remained that way when Louis XIV died.

When Louis the XIV died, his successor at the time Louis XV was too young to rule so Duke D'Orleans was appointed as a regent and was in charge of managing the country's affairs for the too young ruler who was only seven years old. John Law was an associate of Duke D'Orleans through his gambling exploits and used this association to position himself for financial gain.

Duke D'Orleans soon realized after he began managing the country's affairs that France was so deep in debt that it couldn't even cover the interest payments on that debt. John Law had rare mathematical talents and had already published papers on economics. He immediately sensed an opportunity and presented economic papers to his buddy Duke D'Orleans which blamed France woeful economic problems on it having an insufficient amount of currency. John Law touted the use of paper currency as the solution in his presentation.

The Duke bit the bait and on May 17, 1716, gave John Law a bank name Banque Generale and the right to issue paper currency. Its notes would thereafter be used in payment of taxes. Its capital was purchased 25% in coin and 75% in oversupplied state bills at face value which at that time was trading at a heavy discount. Then by guaranteeing his paper money not with just any coin, but with the coin in issue at the time of a note's creation he quickly found his paper was preferred over coinage, which had recently been debased once and was expected to be debased further.

In this way he collected most of the country's stock of precious metal. By the end of the year his bank notes were worth 15% more than equivalent coinage while the state's debts were trading nearly 80% below nominal.

The newly issued paper currency brought some life back to France's economy and John Law was hailed and celebrated for this occurrence and was rewarded by Duke D'Orleans with the rights to all trade in France's Louisiana Territory in America. The Louisiana Territory was vast and stretched from Canada to Mississippi.

This was lucrative for John Law because Louisiana was thought to have gold and his newly created Mississippi company had the exclusive rights to trade in this area and quickly became the richest company in France.

John Law capitalized on the prospects of his company by issuing shares to the public who quickly scooped up these shares with the discounted state bills which no one could wait to get rid of. Law continued to have a succession of stock issues each at higher prices than the previous ones as the monopoly rights to trade with the east was also given to the company. The public rushed to get rid of its increasingly valueless state bills for Law's bank notes.

John Law soon became one of the most powerful financial figures in France and Duke D'Orleans rewarded Law for this feat by making John Law's bank the Banque Royale, the central bank of France. The Duke also gave John Law the sole right to coin and refine all silver and gold and the monopoly on the sale of tobacco.

Now that John Law's bank was the central bank of France this meant that his new paper notes where backed by the government. This created a financial momentum of its' own. The Duke encouraged Law to issue more and more notes. This produced an illusory prosperous time for France.

Charles Mackay in his book summed up these times he stated in it:

"It was now that the frenzy of speculating began to seize upon the nation. Law's bank had effected so much good, that any promises for the future which he thought proper to make were readily believed. The Regent every day conferred new privileges upon the fortunate projector. The bank obtained the monopoly of the sale of tobacco; the sole right of refinage of gold and silver, and was finally erected into the Royal Bank of France.

"Amid the intoxication of success, both Law and the Regent forgot the maxim so loudly proclaimed by the former, that a banker deserved death who made issues of paper without the necessary funds to provide for them. As soon as the bank, from a private, became a public institution, the Regent caused a fabrication of notes to the amount of one thousand millions of livres. This was the first departure from sound principles, and one for which Law is not justly blameable."
- Charles Mackay, 1841.

Nevertheless, things began to unravel for France and John Law. Prince De Conti wanted to buy as much as the Mississippi company's stock as possible at the price that he wanted, but John Law refused and offended the Prince. Upset, the Prince in turn demanded immediate settlement in gold and silver coins to cash in all of his stock in John Law's paper currency.

The Prince was paid with 3 wagons full of gold and silver coins. The Duke got pissed off at this action by the Prince and demanded that he returned the gold and silver coins if he ever wanted to set foot in France again. The Prince complied however he only sent back two of the three wagons full of gold and silver.

By being informed of the Prince's actions, the smart money began a mad dash to get rid of all of John's Law paper currency that they were holding. The professional people at that time started cashing in their paper to buy gold and silver coins, bullion, jewelry and anything of transportable value. They hoarded these valuables or shipped them abroad to Belgium, Holland, and England.

In an attempt to stop this rush to gold and silver by the people of France, John Law with the little authority that he had left abolished the coins as a medium of exchange.

In February of 1720, he declared it illegal to own more than the tiniest sums of gold in any form. He then followed this up by closing the borders and sending instructions to all the coach houses to refuse fresh horses to anyone traveling to foreign lands until an inspector had examined their baggage.

Rewards were offered to the citizens for filing a report on those who had gold and silver in their possession. The crisis got to its' boiling point and on May 27, the banks were closed and John Law was dismissed from his duties. The paper currency that he created was devalued by a whopping 50%.

Once this devaluation took place banks reopened and once again allowed redemption of the bank notes at their new value for gold and silver. And of course a mad rush ensued to convert these highly devalued bank notes. To no one's surprise because of the incredible demand, the banks simply ran out of gold and silver. When they ran out of these precious metals people were paid with copper.

John law was persona non grata in France and relocated to Venice where he died broke. As a result of the failing of the fiat currency that he created, France and the rest of Europe were sunk into a depression which lasted many years.

Whatever Happened To The Classical Gold Standard?

From the time period of 1871-1914 during the beginning of World War I, most of the developed nations of the world which included the Unites States operated under the classical gold standard which simply meant that the currency was pegged to gold.

So each nation's currency for example the dollar, franc, and the pound was merely a name that represented a definite weight in gold. For example, the dollar represented 1/20 of a gold ounce and the British pound sterling also represented a fixed value.

The currencies were pegged to each other so when the people and nations contracted business, they knew exactly what the exchange rate would be because it was fixed. The advantages for developed nations under the classical gold standard were that the currency issued was actually real, because it was backed by the gold and silver in the treasury and on average there was no inflation.

I mean there were a few inflationary and deflationary periods here and there, but over time it all averaged out to zero because everything was based on gold. So what happened to the classical gold standard and why did it die?

The first part of that answer is that the classical gold standard broke down during World War I, when the countries that were in the war resorted to inflationary finance to fund it. And for the second part of that answer you have to look at the creation of the Federal Reserve in 1914 and its true aims.

Was it advantageous to the Federal Reserve that the United States remained on the classical gold standard or on any gold standard for that matter? The answer is no, because they wouldn't have been able to issue its monopoly money that is back by nothing and run their shenanigans through the fractional reserve lending system which I previously discussed earlier. Now let's look at the Federal Reserve and how it came into existence.

How The Federal Reserve Was Manipulated Into Existence

In 1907, JP Morgan an agent of the Rothschilds who not so coincidently face appears on the game monopoly published a rumor that a major bank in New York was insolvent. This created mass hysteria and led people to start withdrawing their money from the banks causing a domino effect. This was known as the Panic of 1907. Banks began calling loans which forced many people to sell their properties to meet loan obligations. Many didn't have the cash to meet these sudden calls on loans which led to an enormous amount of bankruptcies and overall economic turmoil.

The panic of 1907 was mainly engineered by JP Morgan to make way for the creation of the central bank in the United States. As a result of the panic, Senator Nelson Aldrich who later became related to the Rockefellers by marriage recommended that a central bank should be implemented so that the panic of 1907 would never happen again.

In 1910, a secret meeting took place on Jekyll Island which was located off the coast of Georgia. This secret meeting was attended by Paul Warburg one of the wealthiest men in the world at that time and also an agent of the Rothschilds and six others who in total represented 1/4 of the wealth of the entire world to write the Federal Reserve Act.

Of course this act was designed and written by you guessed it the international bankers. The bankers used Senator Nelson Aldrich to push the Federal Reserve Act through congress two days before Christmas when the majority of Congress was already on their holiday vacation, so it easily passed without opposition.

In 1913, the Federal Reserve Act was signed into law by President Woodrow Wilson and the bankers basically hijacked the money supply as the United States government relinquished its right to coin money and regulate its value thereof, giving that absolute power right to the Federal Reserve.

Ever since the Federal Reserve took over the money supply the dollar has lost 95% of its value.

Thomas Jefferson the 3rd President of the United States foresaw the potential damage and effect that a central bank would have on citizens as he stated:

"If the American people ever allow private banks to control the issue of their money, first by inflation and then by deflation, the banks and corporations that will grow up around them (around the banks), will deprive the people of their property until their children will wake up homeless on the continent their fathers conquered."

He further added:

"I believe that banking institutions are more dangerous to our liberties than standing armies."

Congressman Louis T. McFadden who, for more than ten years, served as chairman of the Banking and Currency Committee, stated that the international bankers are a "dark crew of financial pirates who would cut a man's throat to get a dollar out of his pocket... They prey upon the people of these United States."

John F. Hylan, the mayor of New York, said in 1911 that: "The real menace of our republic is the invisible government which, like a giant octopus, sprawls its slimy length over our city, state and nation. At the head is a small group of banking houses, generally referred to as "international bankers".

Most recently Congressman Ron Paul of Texas gave his view on the Federal Reserve Act and the creation of the Federal Reserve. He said:

"I consider the Federal Reserve Act and the creation of the Federal Reserve as being unconstitutional. It gave the power to the Federal Reserve to create money out of thin air. The notion of a central bank doesn't fit in the constitution. The congress has the only authority to coin money and only gold and silver should be legal tender. It is an absolute contradiction to the constitution to have a Federal Reserve system and a central bank".

You heard the quotes from some of the people who opposed a private banking system that is above the law, not subjected to audits and answers to no one. Now let's hear it from the perspective of one of the original international bankers Mayer Amschel Rothschild who sent five of his sons to the different parts of Europe (Frankfurt, Naples, London, Paris, and Vienna) to establish central banks in order to gain control of those nations money supply. Mayer Amschel Rothschild stated:

"Permit me to issue and control the money of a nation and I care not who writes its laws"

So there you have it, the history and the opposition to the creation of the Federal Reserve that controls the present day banking system in the United States. What I highlighted is not based on a so-called conspiracy theory, but it is entirely based on historical facts.

The International Bankers Fund Both Sides In A War

Many people don't know this but the international bankers fund both sides of a war because it is good business for them to do so. They consider it good business because war is a costly endeavor and most nations can't just finance a war through the collection of taxes from its citizens so they have to borrow the money.

They borrow this money from the international bankers and have to pay them back both the principal and interest which are hefty. So in essence, in order to borrow this enormous amount of money it takes to fund a war a nation's government has to surrender a measure of its sovereignty as collateral such as its natural resources and control of its money supply.

Examples of how the international bankers funded both sides of a war is when during the Civil War in the United States the North was financed by the Rothschilds through their American agent August Belmont and the South was financed through the

Erlangers who were the Rothschilds' relatives.

In addition, during World War I the German Rothschilds bankers loaned money to the Germans, the British Rothschilds bankers loan money to the British, and the French Rothschilds bankers loaned money to the French.

Speaking on the subject of World War I, as with the case of wars throughout history anytime there is a war the countries that were fighting in it suspended the rights of its citizens to redeem their gold and the battling countries also got into a tremendous amount of debt as a result of the war.

For example, at the start of World War I in 1916, the United States national debt was $1 billion. It then rose to a peak of $26 billion in 1919 to finance the war. Their solution to this dilemma at the time as it has always been throughout history in the case of war was to print more money and inflate the currency.

The Fake Gold Standard

After World War I, countries around the world wanted to return to the stability and great trade that the classical gold standard era brought them before the war. But they didn't want to reduce or devalue their inflated currency against the gold that they had by matching the amount of currency to that gold. So they decided to return to a watered down version of the classical gold standard. This new fake gold standard was called the gold exchange standard.

Since the United States owned most of the gold in the world after World War I, it was decided by the powers that be that the U.S. dollar and the British pound along with gold under the gold exchange standard would be used as currency reserves by the Central banks throughout the world and in addition, the British pound and the U.S. dollar would be redeemable in gold.

But as you know already the Federal Reserve was created in 1913 and had the ability to create money out of thin air. So how were they able to print this money and still have all the gold to back it up? That answer can be found in the Federal Reserve act of 1913, which stated that the Federal Reserve must provide a reserve in gold equal to 40 percent against the Federal Reserve notes in actual circulation.

Here's the catch though, with fractional reserve lending added in to the mix this became nothing more than a pyramid scheme with the Fed Reserve on top, because once the money that they've created gets deposited in the banking system those banks were only required to keep 10% as a reserve and they were allowed to use the remaining amount as a basis for new loans, thus expanding the money supply even further and the process just simply repeated itself as other banks entered into the equation.

During the 1920's, there was an abundance of money created out of thin air by this banking pyramid scheme. In fact, the Federal Reserve increased the money supply by more than 61% in three years. This enabled commercial banks to loan out money to people and businesses. Interest rates were very low so money was cheap to borrow. This era was known as the roaring 20's and was characterized by a worry free attitude by the American public.

The Roaring 20's Boom Then Bust

Stock market speculation was rampant during this time period and people stopped saving their money. As the Federal Reserve kept interest rates low bubbles emerged in the real estate and the stock market. The stock market rose and investors piled in, borrowing money to cash in on the bubble.

Stock prices rose far faster than corporate profits and stock values on paper far exceeded what companies were worth. In 1928, the market went up by 50% in just 12 months. Things were booming and the American public was drunk on prosperity. However, this was a great big set up and would soon end in a devastating and crippling fashion. I referred to this as the great big set up because all the Wall street giants of that era like John D. Rockefeller and J.P. Morgan divested from the stock market and put all their assets in cash and gold. Did they know something? You be the judge.

In 1929, the bubble burst and the stock market crashed. By 1932 and 1933 the stock market declined by a whopping 80%. The economy stalled as a result of people becoming poorer as a result of their stock market losses. The demand for goods declined dramatically.

To compound this dire financial situation even further, the banks tried to collect on the loans made to stock market investors who bought stocks on margin, but their holdings were now worth little or nothing. When the public realized that the bank assets contained huge uncollectable loans and worthless stock certificates bank runs occurred.

People rushed to take their money out of the banks however, the banks didn't have the cash to back up those deposits. Banks began failing by the thousands and back in those days there wasn't any FDIC Insurance to protect the depositors so they lost everything.

The Great Depression

Simultaneously, while this was occurring Great Britain defaulted from the gold exchange standard and devalued the pound. Cutting the link between the amount of gold and the amount of money. International investors as well as other countries thought the United States would default from the gold exchange standard also so they began to redeem their dollars for gold. The Great Depression ensued and gold began exiting out of the country to foreign lands.

During the depression in 1933 the quantity of money in the United States had decrease by 1/3. For every $3 of currency in deposits that people had in 1929 only $2 was left. For every 3 banks that were open in 1929 in 1933 only two were left. So in actuality the banking system process was working in reverse because banks were destroying money.

Well technically they weren't really destroying money, because remember this money was being created out of thin air. So what really happened was the fractional reserve lending process was thrown into chaos because the way it was structured to work was centered on the total amount that the bank was required to keep as its reserves against its deposits.

All in all for this process to work smoothly for the banks and the fractional reserve system, the banks counted on the deposits that were being made by its customer to average out to equal the withdrawals made.

This didn't happen when bank runs ensued because withdrawals started exceeding deposits and the banks had to draw on those excessive reserves which quickly became exhausted. When those excessive reserves were gone the banks had to liquidate the loans that it had made. The banks had to call the loans due and this resulted in the severe contraction of the money supply.

Bank Holiday

When President Franklin Delano Roosevelt took office in March of 1933 to stem the massive bank runs that were occurring and flow of gold exiting the country issued an executive proclamation closing all banks for a "bank holiday" which lasted for three days. In April of 1933 he also issued an Executive Order banning the private ownership of gold coins and bullion.

The citizens of the United States were forced to surrender all of their gold to the Federal Reserve in exchange for Federal Reserve notes. Not only did FDR ban private ownership of gold, his order threatened fines and imprisonment for any American who failed to surrender his or her gold for paper money. Roosevelt also signed the Thomas Amendment that effectively decreased the gold content of the dollar to 40.94 percent.

But Roosevelt wasn't finished yet. During this time most contracts contained a "gold clause" which was basically designed to protect investors from currency inflation. This "gold clause" stated that the investor was allowed the right to be repaid in gold coins.

However, Roosevelt influenced congress to pass what was known as the Gold Clause Resolution of 1933 which prohibited any contract which gave one of the parties the option of requiring payment in a specified amount of gold.

This affected all contracts that contained the "gold clause" including government contracts. In fact, most of the government bonds contained "gold clauses". So essentially the United States government stole from the people.

This theft continued when Roosevelt signed a proclamation that devalued the dollar. Before this devaluation took place it took $20.67 to exchange for one troy ounce of gold. Now it took $35 in exchange for that same ounce.

Most people knew that this was highway robbery by the U.S. government and didn't turn in their gold. Only 22% of the populace complied and those who made the decision to hang on to their gold realized an almost 70% profit.

In the end, Roosevelt's devaluation of the dollar, along with not allowing claims to gold by United States citizens and increasing the requirement that it took for foreign central banks to purchase units of gold saved the fractional reserve banking system.

The devaluation of the U.S. currency also had the effect of stifling an international run on the dollar. This kept international trade afloat because in essence the United States was operating on a gold standard as a result of the revaluation of gold. The value of gold now matched the amount of value of the monetary base. Although gold won out as history proves time and time the people lost as their wealth was confiscated.

Did FDR's New Deal Get Us Out Of The Depression?

Many people believe that FDR's New Deal and the related public works program lifted the United States out of the Great Depression but it did not. The New Deal actually failed. Instead of it creating growth in the private sector it created big government spending and growth that squeezed the private sector through taxation.

It supposedly created jobs for the vast number of Americans that were jobless at the time, but to create those jobs the money had to come from taxing the citizens. So in actuality the new deal stole a portion of the wealth from some and simply gave it to others. Unemployment during this time still remained high at 18% in spite of this so-called job creation. The New Deal turned out to be nothing more than smoke and mirrors.

So what got the United States out of the Great Depression? It was a combination of the increased amount of gold coming in from Europe, the threat of Hitler, and the prices of goods remaining the same while the price of gold increased by 70%.

When FDR devalued the currency by 40% it decreased the purchasing power of the dollar in both the United States and abroad. As a result, the United States imported less goods because the dollar bought less. Although the dollar bought less at home foreigners realized that their currency bought more goods in the United States as ever before, so they simply bought more and more goods which stimulated the United States economy out of the depression.

In addition to this, Hitler annexed Austria and the European investors and businesses got scared so they began transferring their wealth to United States investments. The results were that there was an influx of gold in the United States. Combined with the Gold it already had plus what it confiscated from the citizens, the Federal Reserve the central bank of the United States owned a majority of the world's gold supply.

Bretton Woods Agreement

The United States was selling arms to these European nations during World War II and they were paying with their gold. Alarmed that once they got out the war their gold supply would be severely diminished, in July of 1944, they held a conference in Bretton Woods, New Hampshire with all the allied powers in attendance to address those and other issues regarding international trade.

Recognizing that the United States represented half of the global economy, the dollar was made the global reserve currency. All foreign currencies had fixed rates of exchange to the dollar which was redeemable for $35 an ounce of gold. However, they were only redeemable by foreign central banks. The majority of the world pegged their currencies to the United States dollar.

But the main flaw of the Bretton Woods Agreement was that there was nothing in the agreement that prevented the United States from expanding the supply of the Federal Reserve notes. As a result of this the gold backing behind the U.S. dollar steadily declined, because there was not enough gold to back all of the dollars that were being printed.

As the Vietnam War got hectic the Federal Reserve just printed more money because the President at the time Lyndon Johnson refused to pay for the war through taxation. The results were deficit spending and the flooding of the world with paper dollars. The President of France Charles De Gaulle became skeptical that the United States wouldn't be able to honor the Bretton Woods Agreement to redeem their excess dollars into gold.

France began converting their cash to gold and withdrew from the London gold pool which was supposed to be some sort of regulatory outfit where the central banks across the world sold tons of gold into the marketplace to suppress the price of gold and keep it at $35 an ounce.

Great Britain then devalued the pound during this time which caused a run on gold. The effect that this had on the gold pool was instead of selling a few tons a day it was now selling more than 200 tons per day. The gold pool eventually closed as a result of this and gold now in the free market was allowed to set its own price.

Gold was defeating the dollar. President Nixon soon began feeling the constraints of having the dollar on the gold standard, so on August 15, 1971, he closed the gold window by taking the United States off the gold standard. The dollar was no longer convertible to gold.

All currencies throughout the world instantly became fiat currencies because through the Bretton Woods agreement the world's currencies were pegged to the dollar because the dollar was backed by gold. So the Bretton Woods agreement essentially failed.

Gold Made Legal Again!

On December 31, 1974, Gerald Ford made it legal for U.S. citizens to own gold again. U.S. citizens began to immediately buy gold and trade it. The prices of gold started rising. Ever since Richard Nixon took the dollar off the gold standard the price of gold rose from $35 per ounce to over $200 per ounce in 1978. What drove the price up of gold dramatically was that people began speculating and there were those who simply bought gold as a hedge against inflation.

People began to line up in droves at their local coin shops to purchase this precious metal. America had definitely developed gold fever and this drove up the price of gold even further to more than $300 an ounce in 1979.

Why did gold jump up in value at least 10 times since it left the dollar? Simply because once gold left the dollar and was allowed to be traded freely gold revalued itself resulting in a transfer of wealth. You must understand that wealth is never destroyed it simply transfers itself. For example, if you were to have bought gold in this period when it was $50 an ounce all you had to was hold on to it as the price rose to $300.

Once the price rose to $300 as a result of people rushing to buy gold all you had to do was sell your gold to them at that $300 price and you would have increased your wealth six times. Those whose simply bought at the $300 price simply transferred their wealth to you.

The bottom line is at the point that I am writing this book gold is roughly around $930 an ounce and silver is around $20 an ounce. As the Federal Reserve injects more and more money into the economy by bailing out banks and big businesses and the government continues to spend trillions on the so-called economic stimulus which further debases the dollar I predict gold will reach $3,000 to $5,000 an ounce soon and its counterpart silver will rise as well.

Actually silver will turn out to be the better investment than gold because it is extremely undervalued, so its upside is tremendous. My question to you is this, are you going to stand on the sidelines and watch gold and silver prices go through the roof or are you going to get in the game and be part of the biggest wealth transfer in history?

Inflation Comes Roaring Back

Inflation starting kicking in and catching up to the United States in the 1970's as a result of the deficit spending that occurred during the Vietnam War and the funding of social programs. In addition, the United States was hit with an Arab oil embargo for supporting Israel during the Yom Kippur war.

Gas and oil prices skyrocketed and the combining factors just mentioned produced stagflation, an economic condition of continuing inflation, stagnant business activity and an increasing unemployment rate.

Inflation during this time period fed on itself as people began to expect continuous increases in the price of goods so they bought more. This increased demand pushed up prices even more, leading to demands for higher wages which pushed prices even higher in a continuous upward spiral.

The bottom line was in the 1970's the U.S. economy was terrible because of stagflation. To give you a bird's eye view on the effects of inflation let's look at these following examples. A postage stamp in the 1950's cost 3 cents and today it cost 45 cents which amounts to 1,300% of inflation. A house in 1959 cost $14,100 and today's medium price for a house is $213,000 a whopping 1,400% of inflation.

The purchasing power of the dollar has severely diminished because of the constant printing of money and debasing of the currency by the Federal Reserve. An even more clear cut example of this is in 1905 it would cost you roughly $20 to buy a 1 ounce gold coin. For that same 1 ounce gold coin today you will pay at least $900.

The frightening thing is the U.S. dollar is rapidly declining at an accelerated pace and it will eventually crash and most people are not even aware of it as they continue to hold their life savings in U.S. dollars. These people and I hope for your sake you are not one of them will eventually get wiped out.

The Stock Market Crash Of 1987

What preceded the stock market crash of 1987 was an environment characterized as a powerful bull market in the United States and elsewhere across the world. What fueled the bull market during this time were corporate hostile takeovers, leveraged buyouts and mergers. Leveraged buyouts were funded primarily by junk bonds. These junk bonds were highly risky so they paid out at a higher rate of interest. Companies were in a mad dash to raise money to buy each other out which created a frenzy.

Also IPO's (initial public offerings) were commonplace and added fuel to the growing bull market. Computers were also now in vogue and people started viewing the computer as a revolutionary tool that would change their way of life. The market was bubbling, but things began to change in 1987. The S.E.C. conducted numerous investigations involving illegal insider trading which spooked investors.

The strong economic growth that occurred during this time bought on inflationary fears. The Federal

Reserve in response rapidly raised short-term interest rates which hurt stocks.

There were also serious concerns over the trade and budget deficits. Anxiety rose and this led way to a highly volatile market. The sell off began and the Dow began falling simultaneously.

During October 14 – 16 of 1987 the Dow fell over 260 points and the S&P declined 10%. On October 19, 1987 referred to in the annals of history as Black Monday the Dow crashed as a majority of the stockholders tried to sell their stocks simultaneously. The Dow plunged by 508 points which was 22.6% of the market. Chaos was the order of the day as people tried to sell but couldn't because of electronic glitches and because they weren't any buyers to fulfill the orders.

The Federal Reserve in response to the stock market crashing increased the currency supply to stem a potential world-wide economic depression as other parts of the world were also affected. As I explained earlier the Federal Reserve is responsible for creating booms and their accompanying busts. So what happened in this situation when the Federal Reserve increased the money supply was it affected the real estate market in different parts of the country that were booming and turning them into little bubbles and as you already know with bubbles they eventually burst.

The Dotcom Bubble Then Burst

Another great bubble that the Federal Reserve created was the Dotcom bubble of the late 1990's. The Federal Reserve chairman at the time Alan Greenspan lowered interest rates which encouraged venture capitalists to invest their money into internet start-ups.

Basically, during this time anyone with an idea and a business plan that showed the potential of future profits could get funded even if there wasn't any real legitimacy to the numbers. The Dotcom madness fast became a novelty and exploded as the public was swept up and started investing huge amounts of money in these companies through stock purchases.

This led to grossly obscene overvaluations of most of these tech stocks. For example, E-Toys an online toy retailer was considered more valuable than Toys-R-Us. This ultimately became a pyramid scheme requiring new investors armed with new capital to drive up the price of stocks.

The rise of the prices of stocks had nothing to do with sound business fundamentals such as production, earnings, efficiency or profits. It was all based on speculation and the greater fool theory which was as long as a greater fool came along and bought your stock for a higher price you were okay.

I vividly recall that during this time Amazon.com stock price rose to $300 a share despite it having huge losses. So the public essentially became insane which Alan Greenspan referred to as "irrational exuberance". People started believing that these stock prices would continue to go up forever but they were sadly mistaken as the Nasdaq peaked in 2000 to 5049 and subsequently plunged to 1,140 in 2002.

The majority of the Dotcom companies literally fell off the face of the earth and no longer existed. People got wiped out once again, lives were ruined and wealth was transferred. Are you beginning to see the trend here?

The Real Estate Bubble Then Burst And The Sinking Of The Economy

The origins of the housing crisis began as the Federal Reserve once again kept interest rates low and credit was easy. In fact, credit was so easy it produced the whole subprime lending culture where people basically were given predatory loans to buy houses even though they couldn't afford these houses. They were sucked in by the low teaser rates that these subprime mortgages initially offered, not taking into account that these teaser rates would eventually adjust.

What this produced was a housing bubble as real estate prices became insanely inflated as a result of so much liquidity in the market. Everybody had access to money to buy houses and bought houses they did. As the value of these house artificially increased, people began using their houses as ATM machines taking out home equity loans and lines of credit so they could take vacations, buy plasma T.V's, SUV's etc.

Then the subprime sector which accounted for $600 billion or 20% of all mortgages in 2006 started to implode as these mortgages began to adjust at significantly higher interest rates and people were forced to pay higher mortgage payments and basically couldn't. This led to a tremendous amount of defaults and the housing market began to unravel.

There was a whole other industry tied to this fiasco as mortgages and subprime mortgages were sold off to the secondary market which repackaged and securitized them creating what is known as mortgage backed securities.

These mortgage back securities were then repackaged as derivative securities called collateralized debt obligations or CDO's and given investment grade bond ratings. These CDO's were sold to banks, insurance groups, hedge funds, and other institutions who bought them for their high paying yields.

Most of these CDO's were purchased by these financial institutions with borrowed money. So they were in highly dangerous leveraged positions. As housing defaults rose and housing prices fell these institutions began to get pulverized with these risky investments and bad bets and the whole economy began its downward spiral.

A lot of these financial institutions began to sell these bad CDO's to the marketplace at fire sale prices. To add to their misery under an accounting rule called mark to market they were forced to write down the value of the CDO's they had on their books to what they could sell them for at current market prices. This resulted in massive losses on their books.

For banks this had a tremendous effect on how much money they could lend to customers as they are required to maintain a certain percentage of equity to the amount of loans they lend. This equity ratio turned adversely negative and many banks could no longer loan money so they began going out of business.

And on top of that Fannie Mae and Freddie Mac who hold nearly half of the United States mortgages had to be bailed out by the government because they had 5 trillion dollar of mortgage risks exposure but only $80 billion in capital. Everything was so intertwined. Hence, the economic meltdown of the whole economy ensued with a devastating credit crises in the United States and globally.

So that's where we are at today and the problems are getting even deeper.

The National Debt

The official debt of the United States as indicated by the government is 10 trillion dollars. This is a total lie because this figure ignores Social Security, Medicare and the new prescription drug benefit.

According to Richard W. Fisher President of the Dallas Federal Reserve the United States national debt is closer to 100 trillion dollars. He states in a May 28 2008 speech:

"Add together the unfunded liabilities from Medicare and Social Security, and it comes to $99.2 trillion over the infinite horizon. Traditional Medicare composes about 69 percent, the new drug benefit roughly 17 percent and Social Security the remaining 14 percent."

The $99.2 trillion dollars that Mr. Fisher mentions doesn't include other unfunded liabilities and the future cost of wars. To understand what it would take mathematically to pay this whole $99.2 trillion amount off he states in his speech:

"Let's say you and I and Bruce Ericson and every U.S. citizen who is alive today decided to fully address this unfunded liability through lump-sum payments from our own pocketbooks, so that all of us and all future generations could be secure in the knowledge that we and they would receive promised benefits in perpetuity.

How much would we have to pay if we split the tab? Again, the math is painful. With a total population of 304 million, from infants to the elderly, the per-person payment to the federal treasury would come to $330,000. This comes to $1.3 million per family of four—over 25 times the average household's income."

The reality of this situation is my friend is that the United States of America is flat out broke. I know that this hard to swallow for some but it's the truth. That's why it's hard to understand and flat out ridiculous how President Obama and his administration can champion more spending as a solution to America's financial problems.

That's like giving a drunk more alcohol as a solution to their alcoholism or adding more gasoline to the flames as a solution to putting out a roaring fire. It is simply ludicrous and moronic.

Add in the massive government spending that is taking place right now, the enormous trade deficit, the massive economic stimulus and bailouts and the United States is sinking deeper into the abyss of no return. Trillions and trillions of dollars are being spent and somebody's got to pay for all of this spending. If you are a United States citizen that somebody will be you in the form of taxes and inflation.

However, this time around unlike the great depression of the 1930's which was a deflationary depression (meaning prices actually went down) the Greatest Depression that we are currently experiencing right now in its earliest stages as we move further along in it will eventually turn into an inflationary depression and possible hyperinflationary depression like the one's experienced in Weimar Germany in the past and what's happening in Zimbabwe today.

How soon this happens is anyone's guess, but it will happen as the dollar continues to decline further. We are already experiencing rumblings from countries like China who hold most of our debt saying that they know that the United States will not be able to pay them back and will eventually default on its payments. They are even pushing to get rid of the U.S. dollar as the world's reserve currency and will eventually unpeg its currency the Yuan from the U.S. dollar.

When these scenarios unfold look for other nations to follow suit and this is when the dollar will come crashing down and become worthless. The question I primarily get asked when I make these strong predictions is won't the Federal Reserve pull out all stops to save the dollar? My answer is always this.

The Fed has already tipped its' hand as to what it is going to do. The present Federal Reserve chairman Ben Bernanke a student of the great depression said in a speech made years ago that he would "drop money from a helicopter" to avoid deflation. He further stated, "people know that inflation erodes the real value of the government's debt and therefore, that it is in the interest of the government to create some inflation."

There you have it straight from helicopter Ben's mouth. The bottom line is you have to protect yourself and your wealth from such a destructive monetary policy. How do you protect yourself? You must have a hedge against inflation. Gold and Silver are that perfect hedge against inflation and a declining dollar.

To Save or To Hedge

To position yourself to offset inflation and the declining dollar you must know why if you save money in a bank or under a mattress you are inevitably losing. I know that growing up as a child you were probably taught that saving money was good so let's look at critically what you were taught.

When you save money in a bank and get paid interest on your savings first of all the government taxes that interest. So in effect you are actually being penalized for saving. In addition, the interest rates are so low at the bank that the interest that they do pay you doesn't keep up with the rate of inflation so you are actually losing money by keeping it in the bank.

The same principle applies with keeping it under a mattress, you still lose. It's not moving or growing so therefore it will eventually die that's why they call it currency. The bottom line is, saving money doesn't suffice in the type of world we live in today. It's an old school concept that is obsolete and can leave you wiped if you don't adjust your thinking and adapt to the concept of hedging.

So what is a hedge? A hedge is a means of protection or defense, especially against financial loss. So what are we trying to protect? You are trying to protect your purchasing power, you are trying to protect your wealth. You don't want it to be decreased or be wiped out. So you hedge against this possibility. Hedging is basically a defensive move.

So that's one of the main reasons why you invest in gold and silver. You invest in these two precious metals as a hedge against inflation and a declining dollar. Here are some facts pertaining to gold and its use as a hedge against inflation.

1) As inflation goes up the price of gold goes up with it.

2) Since the end of World War II, the five years in which U.S. inflation was at its highest in the United States were 1946, 1974, 1975, 1979, and 1980. During those five years, the average real return on stocks, as measured by the Dow, was

12.33%; the average real return on gold was 130.4%.

3) Gold is real money and it's one thing that the banksters can't print into oblivion unlike paper currency which they can inflate at will. When they inflate the currency prices rise and the purchasing power of the dollar diminishes.

In fact, it can go so low that merchants may not want to accept it anymore as a medium of exchange like what's happening in Zimbabwe where its citizens are forced to pan for gold in order to buy food because shopkeepers are not accepting Zimbabwe dollars.

The last time I checked a loaf of bread in Zimbabwe costs 1.6 trillion dollars and as recently as August 2008 Zimbabwe was experiencing an inflation rate of 11.2 million percent. How do you think the people made out who bought gold as hedge when the currency wasn't so bad?

Silver is also a great hedge against inflation and financial turmoil. It also goes up as inflation goes up. Historically, in previous periods of financial turmoil and high inflation silver has always out performed gold by a factor of more than two.

Also silver provides a great hedge because at the present moment it is highly undervalued. Take into the account the gold/silver ratio which is defined as the number of ounces of silver it takes to buy one ounce of gold. Historically the Gold/Silver ratio was 15:1. The current ratio is 72:1 meaning that it takes 72 ounces of silver to buy one ounce of gold. This means that based on historical figures silver is highly undervalued.

There is also another reason to be highly optimistic about the price of silver rising. Silver is an industrial commodity and is consumed in vast quantities by many industrial processes and is a vital material for electronic devices like mobile phones and personal computers. So it is being used up and will continue to be used up and the worldwide supply of it will decrease because the mining of it is severely down.

This makes silver a scarce resource with unique properties as both a conductor of electricity and a traditional unit of currency. Silver has a double hedge against inflation as both an industrial commodity and a precious metal.

So in essence, you can say in investment terms that I am long on gold and silver meaning that I am holding them for the long term because I believe their prices will continue to soar while I am also shorting the dollar which basically means that I am betting against it and believe it will sharply decline in value.

Silver and Gold Price Manipulation

The price of gold and silver are manipulated to intentionally suppress their prices. You have to understand that the central banks create and supply the world with fiat money and gold and silver which are real money is an enemy to this funny money that they issue.

It is an enemy because gold and silver are competitive currencies and their value greatly influences interest rates which ordinarily governments like to keep low.

Evidence that gold and silver prices are manipulated is in August of 2008 two banks held more than 60 percent of the short positions in silver on the New York Commodities Exchange and in addition just three or fewer banks now hold half the short positions in gold on the Comex and more than 80 percent of the silver short positions.

What does all this mean? It means that all of this gold and silver that the banks are promising to deliver at a certain date doesn't really exist in reality. It exists only on paper and this paper is known as a futures contract. How do I know that this gold and silver really doesn't exist? Because they are promising to deliver more silver and gold that is known to be in existence.

In reality, if the traders who are long meaning they are the ones who bought this gold and silver from the shorts through these futures contracts demanded actual delivery there would be a major default and the whole scheme would unravel.

Price manipulation also occurs through silver leasing, silver certificates and through Exchange-Traded Funds (ETFS). The bottom line is this manipulation will soon blow up thereby exposing itself. When this happens look for gold and silver prices to go through the roof.

How To Invest In Gold and Silver

There are several different ways to invest in gold and silver. This section will highlight the various ways in which to invest.

1) **Physical Ownership** – For me this is the best way to own gold or silver because you have actual possession of it, that's if you store it in your home or some other convenient location.

One of the main purposes of having your gold and silver at arms length is that you have immediate access to your wealth without any type of interference. Case in point, when 9-11 occurred my bank located in New York City was closed for three days and the ATMs were down as well. Lucky for me the local coin shop was open and I was able to take my gold and silver coins and sell them for cash.

Another reason for having gold and silver in your possession is that having all of your money in a bank is a risky proposition because you may not be able to get it all out at once which was the case in the 1930's when Roosevelt ordered a bank holiday to slow down the bank runs that were occurring.

You can physically own gold and silver by purchasing bullion which refers to precious metals in their bulk form. This can be gold and silver in the form of ingots, bars or sometimes coins such as the American Eagle, South African Krugerrand, the Canadian Maples leafs and others. I own both bars and coins and I don't have a specific preference. The coins that I primarily buy are the American Silver Eagles and the American Gold Eagles. The American Silver Eagle coins were first minted in 1986 by the United States government.

They are made of 99.99% silver and measure 40.60mm in diameter and contain 1 troy ounce of silver. The silver eagle edges are reeded and they contain a total of 201 reeds. The American Silver and Gold Eagle coins are guaranteed by the Unites States government for their weight content and purity. The silver and gold eagles are legal tender and have nominal values but the reason why you invest in them is for their bullion value which is determined by mass and purity resulting in much higher values.

The US Mint has offered Silver Eagles in three primary versions. The first is the regular uncirculated bullion version of the coin which is available from coin dealers or authorized bullion dealers.

The second is the collectible proof version available directly from the US Mint. The final and most recent version is a collected uncirculated version of the coin also known as the Burnished Silver Eagle. These coins carry the "W" mint mark and are struck on burnished blanks. I'm not a coin collector so these collectible versions of the coin don't matter to me.

The American gold eagle coin was also minted in 1986 by the United States government and they are offered in 1/10 oz, 1/4 oz, 1/2 oz, and 1 oz denominations.

The one-ounce gold American Eagle has a diameter of 32.7mm, a thickness of 2.87mm, a total weight of 1.0909 troy ounces contains one troy ounce of pure gold, and has a face value of $50.

The half-ounce gold coin has a diameter of 27mm, a total weight of .5454 troy ounces, contains a half-ounce of pure gold, and has a face value of $25.

The quarter-ounce gold coin has a diameter of 22mm, a total weight of .2727 troy ounces, contains a quarter-ounce of pure gold, and has a face value of $10.

The tenth-ounce gold coin has a diameter of 16.5mm, a total weight of 0.1091 troy ounces (3.393 grams), contains a tenth-ounce of pure gold and has a face value of $5.

All American Eagle gold bullion coins are 22 karat gold, and contain an alloy of silver and copper to help increase the stability and scratch-resistance of the coins.

The gold American Eagles as well as the price for the American silver eagles are determined by their spot price which is the daily cash price of the metals. When you buy bullion (bars, coins) you will buy it from dealers and dealers basically charge a markup over spot prices. So it is imperative that you do your due diligence and shop around to compare prices.

Here is a list of a few reputable dealers:

American Precious Metals Exchange (http://www.apmex.com)
Kitco (http://www.kitco.com)
Blanchard (http://www.blanchardonline.com)
American Gold Exchange (http://www.amergold.com)

I also buy silver and gold on Ebay and my experience has been great. The thing that I like about Ebay is that a lot of the sellers offer free shipping and you can a lot of times get gold and silver at cheaper rates than the dealers offer. In addition to that the shipping is fast because of the simple fact that Ebay sellers depend on great feedback to stay in business on Ebay so they are eager to please buyers.

Another thing that I need to mention are junk silver coins. Junk silver coins are referred to as junk silver by collectors because they don't have collector's value. However, the reason why you buy junk silver coins are for their investment value. They contain 90% silver and are sold in bags containing $1,000 in face value of dimes, quarters, half-dollars or silver dollars minted before 1968. I believe the chances of these coins being confiscated are slim.

I.R.S. Reporting Requirements

When you purchase gold and silver there are no reporting requirements or any limits to the amount of gold and silver you can own. However, if you buy gold and silver in amounts of $10,000 or more the dealer must report the cash to the I.R.S. Remember, the dealer has to report the cash, not the gold and silver.

When you sell your gold and silver, your dealer must report the sale on form 1099B if:

You are selling 1 or more 1-kilo gold bars, 1 or more 100 ounce gold bars, 25 or more 1-ounce gold maple leafs, Mexican onzas or South African Krugerrands.

In addition, any size silver bar totaling 1,000 or more troy ounces as well as 90 percent silver coins totaling $1,000 or more in face value have to be reported. The great thing is that dealer reporting requirements don't apply to American gold and silver eagles no matter how much you sell.

However, your dealer reporting requirements shouldn't be confused with your own reporting requirements in regards to capital gains. The bottom line is you should seek the advice of a competent CPA or accountant.

2) **Exchange-Trade Funds**- Another way to invest in gold and silver without having to physically store it yourself is through a precious metals ETF. ETF is an acronym that stands for "exchange-traded fund". It is a security that basically trades like a stock and can be purchased and liquidated through any trading platform.

ETFs track indices such as the S&P 500, and The Dow or it can be designed to track a specific commodity like gold and silver. ETF's are cool for trading but for investing I have a bit of problem with them. The problem that I have with them for example like the Gold ETF's is that you never know what's happening with your gold. You don't really know if there is any gold really allocated to you.

Also, when you read the prospectus of the silver ETF Ishares it states that "The IShares are intended to constitute a simple and cost-effective means of making an investment similar to an investment in silver." Basically what they are saying here is that all your shares may not be backed by silver.

The prospectus further reads "the liquidity of the IShares may decline and the price of the IShares may fluctuate independently of the price of silver and may fall". I mean does that sound like you are investing in silver? They're basically saying if silver is rising your shares that you have invested which supposed to represent silver could fall. How could that be so? Sounds a little fishy to me.

The whole point in investing in gold and silver is to have access to it and protect your wealth in uncertain times. In my analysis investing in ETF's defeat this purpose.

3) Buying gold or silver certificates through Perth Mint

The Perth Mint is located and owned by the government of Western Australia and The Perth Mint Certificate Program allows customers to purchase precious metals without the burden of having to personally store them. All accounts are fully guaranteed and insured by the Lloyds of London.

Perth Mint is the only government backed bullion storage facility in the world and investors can purchase bullion silver, gold and platinum at the Perth Mint spot market ask price with no markup.

The minimum investment fee is $10,000 and there is a small service fee of 2.25% and a $50 administrative fee. There is no storage fee and when you want to take possession of your silver or gold you have the choice of having it fabricated into bars and coins. You have to pay to have this done.

Unlike the United States, Australia has no history of gold confiscation, so this may be a good bet if you fear confiscation in the United States. In addition to this, the Perth Mint Certificate program is not a bank account so you don't have to disclose it.

I think that this is excellent for privacy purposes and it gives you a peace of mind in the event that shit hits the fan here in the states at least you know that if they start confiscating gold and silver you have your real money tucked away in a foreign land.

4) **Commodities Futures and Options** – You can invest in gold and silver through commodity futures and options. What are commodities? Commodities are a physical substance, such as food, grains, and metals, which is interchangeable with another product of the same type, and which investors buy or sell, usually through futures contracts. Gold and silver futures are traded on the Comex.

The price of the commodity such as gold and silver is subject to supply and demand and futures and options are types of leverage. A future is basically a contract to deliver a specific commodity like gold or silver in an agreed upon quantity and price at an agreed upon date in the future.

When you enter in a futures contract, you are betting that the value or price of that asset in our case gold or silver is going to be at a certain value at a predetermined time in the future. At that time, when the contact is completed and 'settlement date' arrives, you or the other party cough up with the difference between what was originally paid and what the settlement price is.

Both parties of a futures contract must fulfill the contract on the settlement date. The seller then delivers the commodity to the buyer, or, more often than not, it is a cash-settled future, and cash is transferred from the futures trader who sustained a loss to the one who made a profit.

We are currently experiencing a commodity bull market and trading futures are potentially the most profitable way to participate because you get the full benefit of price increases, net of transaction and carrying costs. And like I mentioned earlier you can use leverage if you know what you are doing to make money.

You can use leverage because you do not have to put up all the money needed for the contract but usually only a small percentage roughly around 10%. So if you enter in a $1,000 contract you only have to put down $100. But if you lose on your prediction you have to cough up the entire $1,000. However, if you are right with your prediction it only cost you $100 to play the game and you made a profit.

There are 4 ways to own future contacts

1) Individual Account- you can open an individual account directly with a registered futures commission merchant(FCM) or you can go through an introducing broker which is a firm that's in the business of soliciting or accepting future orders which are then handled by an FCM. With an individual account all trading and decision making is in your hands.

2) Managed Account - This is also known as a discretionary individual account. In this type of an account you give an account manager the discretion to buy and sell for you. If you choose to go with a managed account make sure the account manager is inline with your investment views.

3) Commodity Pool - In a commodity pool you are a limited partner in a pool of accounts that are managed as one. Your profits or losses are based in proportion to the amount of money in you have invested in the pool. Since you are a limited partner in the pool you have limited liability meaning that you stand to

lose no more what than what you invested.

4) Commodity Index funds – This is a cheap and sensible way to invest passively in the commodities sector and many of its subsectors representing specific commodities. Historically, commodity Index funds have outperformed more than two-thirds of the managed funds. A real popular one is the Rogers International Commodity Index created by famed investor Jim Rogers.

Mining stocks

Another way that you can invest in gold and silver is through owning mining stocks. Investing in mining stocks can be a little tricky because of the fact that mining is capital intensive. In fact, the price of the metals has to rise faster than the cost of producing it in order for miners to attract investors. That's why a lot of mining companies went out of business simply because the cost of mining is expensive and there was no way to make profits.

But once this scenario reverses itself look for mining stocks to take off.

Conclusion

In conclusion, as of the time I am writing this the United States economy has lost approximately 8 million jobs since 2007 when the recession began. The so-called official unemployment rate stands at 8.7% and I think that the true unemployment rate hovers around 12 to 13% because the government figures are skewered. They omit such things in the count as the people who simply gave up looking for work and those who work part time but would rather be working full time.

Nonetheless, you get the picture it's brutal out there! Companies are going out of business left and right. For some reason a lot of pundits and newscasters are still calling this a recession. Some of them are even championing that we have entered in a recovery phase. However, this is pure propaganda because all economic indicators are reading that we are in the beginnings of the greatest depression the world has ever seen.

This depression will be unlike the one in the 1930's which was a deflationary depression and prices actually fell. Back then the U.S. was on the gold standard. Things are similar but a lot different this time. The U.S of course isn't on the gold standard anymore and the central bank of the United States the Federal Reserve is printing tons and tons of money. In fact, they are printing so much money that they stopped publishing how much they are actually increasing the money supply by. This will lead to an inflationary depression.

The United States government instead of letting the market itself make the necessary corrections in order to get out of this thing is interfering in a destructive way that is accelerating the greatest depression the world has ever seen instead of stopping it.

The result of this massive spending is that the currency is perversely being debased. In addition to this the dollar is rapidly declining and will eventually crash to zero and all hell will break loose. If you don't act now to protect your wealth you will not financially survive the greatest economic depression the world has ever seen.

In this book you have gained the necessary knowledge and information on how to protect yourself financially by investing in gold and silver. Yes, I know historically the United States has made it illegal to own gold and has outright confiscated it but don't let this deter you in investing in gold and silver. If this confiscation occurs again a black market will emerge where people will be using gold and silver to buy the essentials of life.

I'm not telling you to break the law, but merely suggesting that you use your intelligence to properly assess the situation when and if that time comes. You and your family's well-being depend on it.

In addition, to investing in gold and silver, it would be wise of you to start stocking up on food and other essential items to prepare for what's coming ahead. I predict that we will see more and more civil unrest in the United States like the occupy Wall street movement as people began to feel more and more that the government and big business are screwing them.

Hopefully, this book and its lessons have given you the motivation, incentive and the tools to not only survive the greatest depression the world has ever seen but to thrive in it as the wealth is being transferred.

Other Books Written By Omar Johnson

Search Engine Domination: "The Ultimate Secrets To Increasing Your Website's Visibility And Making a Ton of Cash"

How To Sell Any Product Online: "Secrets of The Killer Sales Letter"

How To Make Money Online: "The Savvy Entrepreneur's Guide to Financial Freedom"

How To Create A Profitable Ezine From Scratch

The Secrets of Finding The Perfect Ghostwriter For Your Book

How To Make A Fortune Using The Public Domain

How To Overcome Your Self-Limiting Beliefs & Achieve Anything You Want

The Secrets Of Making $10,000 on Ebay in 30 Days

Creative Real Estate Investing Strategies And Tips

The Secrets of Finding The Perfect Ghostwriter For Your Book

The Creative Real Estate Marketing Equation: Motivated Sellers + Motivated Buyers = $

How To Market Your Business Online and Offline

How To Promote Market And Sell Your Kindle Book